This book is dedicated to the memory of my daughter,

Kristin.

Library of Congress Cataloging-in-Publication Data

Paddock, Donna J.

MMM Memorizing Multiplication Tables with Mnemonics

ISBN 13: 978-0-6154-2756-0 Donna J Paddock

ISBN 10: 0-615-42756-1

## Introduction

My daughter, Kristin, was diagnosed with a learning disability called, dyslexia. Dyslexia is a learning disorder that manifests itself primarily as a difficulty with reading, spelling, writing and language. Many dyslexic children have difficulty with rote memory and learning the multiplication tables can be challenging for the dyslexic student.

When Kristin was required to learn multiplication tables she struggled, and I struggled along with her. Every evening after school we practiced the multiplication tables using different approaches. We tried flash cards; we sang the facts; we made colorful flash cards and we resorted to writing them over and over and over again. After each session I would quiz Kristin, and she would give me the correct answer. We were confident that we had successfully learned the tables until the next morning. Somehow during the night Kristin lost the ability to remember the tables. This was very frustrating for both of us.

During this time I was completing my Masters Degree in Education and had selected to do a paper on mnemonics. Mnemonics (it is pronounced "neh-mon-ik") is a learning aid that comes in many varieties and aids in memorizing all types of information. An example is an advertising slogan for a feline rescue center that uses the telephone number 1-800-CAT-HELP.

This word association assists in remembering the number and the purpose of the call. This is a mnemonic or learning aid.

A very popular mnemonic device we have all learned in school is the following rhyme:

> "Thirty days hath September, April, June & November,
> All the rest have thirty one,
> But silly old February spoils the fun."

Kristin and I were running out of time and desperate which lead me to develop this method. Success was measured when my daughter after diligently drawing the multiplication table could wake up the next morning and remember the facts. Since that time I have learned that not only do dyslexic children and ADHD children have difficulty learning the tables but children with high IQ's and above normal reading comprehension can also struggle with multiplication tables.

A key to numbers from one to ten is listed on page "v" and an example of how to use the key and symbols is on page "vi." Each multiplication fact is described on a page within the book. The mnemonic key is drawn for the student to follow. The student is assigned to draw the pictures on paper

**or at the bottom of each page. Merely looking at the pictures and reading the "mnemonic code" will not result in remembering the multiplication facts. Note, that it is not important how well the student draws, but it is *crucial that every picture be drawn.***

**The author assumes that a teacher or parent has shown the student the concept and relationship between addition and multiplication. The student should be aware of the following concept before attempting to learn multiplication facts:**

| | |
|---|---|
| 4 +4 + 4 = 12 | 3 + 3 + 3 + 3 = 12 |
| 4 x 3 = 12 | 3 x 4 = 12 |

| | |
|---|---|
| 8 + 8 + 8 + 8 + 8 = 40 | 5 + 5 + 5 + 5 + 5 + 5 + 5 + 5 = 40 |
| 8 x 5 = 40 | 5 x 8 = 40 |

This method, although it is simple, will be effective in learning and remembering the multiplication tables and will assist in building confidence in the student's learning skills. Have fun and enjoy working with this program,

**Memorizing Multiplication Tables with Mnemonics**

# M M M

# Key to Symbols

1 = Sun 

2 = Shoe 

3 = Tree 

4 = Door 

5 = Hive 

6 = Sticks 

7 = Heaven 

8 = Gate 

9 = Sign 

10 = Pin 

**To correctly use this method to learn multiplication facts, it is critical that the student draw the mnemonic pictures; drawing the pictures is the key to learning the multiplication facts quickly. It is NOT important how well the student is able to draw. IT IS VERY IMPORTANT TO DRAW EVERY PICTURE.**

**EXAMPLE: 8 X 4 = 32**

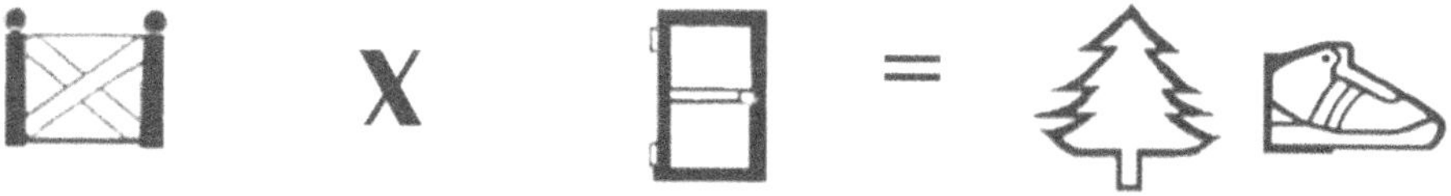

**Student draws 8 gates, 4 doors, 3 trees and 2 shoes.**

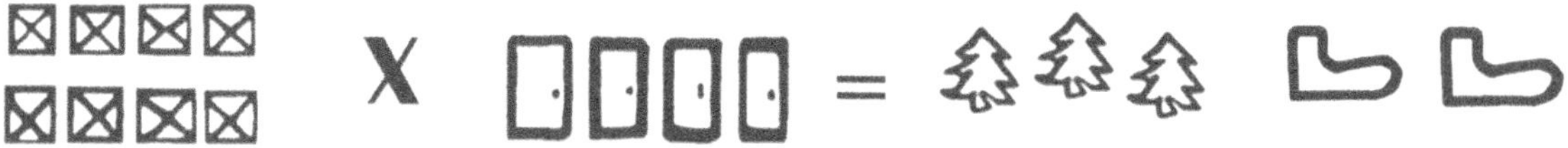

**The student after drawing the pictures states, "Eight gates times four doors equals three trees two shoes. Eight times four equals thirty two."**

**Student writes: 8 x 4 = 32**

**The student may draw the pictures in the book or use a separate sheet of paper to draw the symbols.**

# 1 One

**1 x 0 = 0**

 **x 0 = 0**

**Student draws the following pictures:**

 **x 0 = 0**

**Student states: "One sun times zero equals zero. One times zero equals zero."**

**Student writes: 1 x 0 = 0**

1 x 1 = 1

Student draws the following pictures:

Student states: "One sun times one sun equals one sun. One times one equals one."

Student writes: 1 x 1 = 1

1 x 2 = 2

**Student draws the following pictures:**

**Student states: "One sun times two shoes equals two shoes. One times two equals two."**

**Student writes: 1 x 2 = 2**

1 x 3 = 3

**Student draws the following pictures:**

**Student states: "One sun times three trees equals three trees. One times three equals three."**

**Student writes: 1 x 3 = 3**

1 x 4 = 4

 x 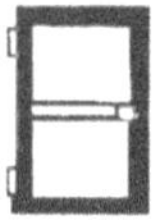 = 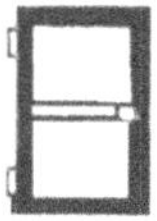

**Student draws the following pictures:**

 x  = 

**Student states: "One sun times four doors equals four doors. One times four equals four."**

**Student writes: 1 x 4 = 4**

**1 x 5 = 5**

**Student draws the following pictures:**

**Student states: "One sun times five hives equals five hives. One times five equals five."**

**Student writes: 1 x 5 = 5**

1 x 6 = 6

**Student draws the following pictures:**

**Student states: "One sun times six sticks equals six sticks. One times six equals six."**

**Student writes: 1 x 6 = 6**

1 x 7 = 7

 x  = 

**Student draws the following pictures:**

 x  = 

**Student states: "One sun times seven heavens equals seven heavens. One times seven equals seven."**

**Student writes: 1 x 7 = 7**

1 x 8 = 8

**Student draws the following pictures:**

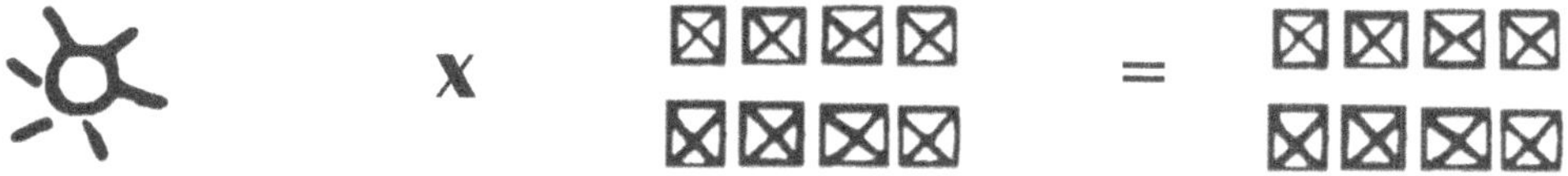

**Student states: "One sun eight gates equals eight gates. One times eight equals eight."**

**Student writes: 1 x 8 = 8**

1 x 9 = 9

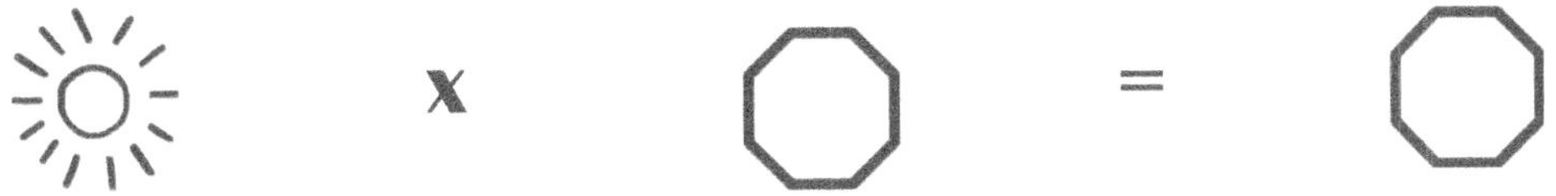

**Student draws the following pictures:**

**Student states: "One sun times nine signs equals nine signs. One times nine equals nine."**

**Student writes: 1 x 9 = 9**

**1 x 10 = 10**

**Student draws the following pictures:**

**Student states: "One sun times ten pins equals ten pins. One times ten equals ten."**

**Student writes: 1 x 10 = 10**

# 2 Two

2 x 0 = 0

 x 0 = 0

Student draws the following pictures:

 x 0 = 0

Student states: "Two shoes times zero equals zero. Two times zero equals zero."

Student writes: 2 x 0 = 0

2 x 1 = 2

 x  = 

**Student draws the following pictures:**

 x  = 

**Student states: "Two shoes times one sun equals two shoes. Two times one equals two."**

**Student writes: 2 x 1 = 2**

2 x 2 = 4

**Student draws the following pictures:**

**Student states: “Two shoes times two shoes equals four doors. Two times two equals four.”**

**Student writes: 2 x 2 = 4**

2 x 3 = 6

**Student draws the following pictures:**

**Student states: "Two shoes times three trees equals six sticks. Two times three equals six."**

**Student writes: 2 x 3 = 6**

2 x 4 = 8

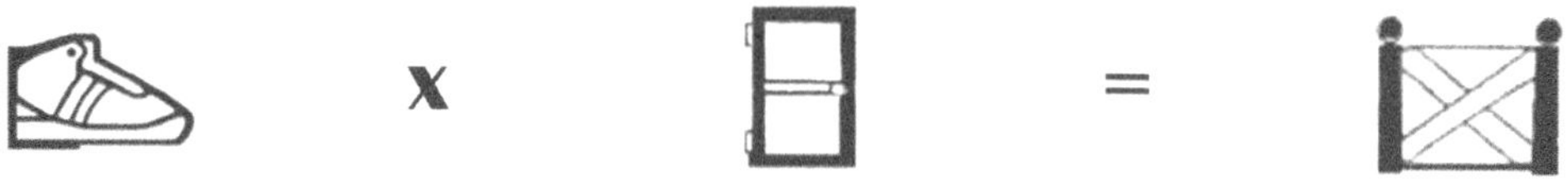

**Student draws the following pictures:**

**Student states: "Two shoes times four doors equals eight gates. Two times four equals eight."**

**Student writes: 2 x 4 = 8**

2 x 5 = 10

 x  = 

**Student draws the following pictures:**

 x  = 

**Student states: "Two shoes times five hives equals ten pins. Two times five equals ten."**

**Student writes: 2 x 5 = 10**

2 x 6 = 12

**Student draws the following pictures:**

**Student states: "Two shoes times six sticks equals one sun two shoes. Two times six equals twelve."**

**Student writes: 2 x 6 = 12**

2 x 7 = 14

**Student draws the following pictures:**

**Student states: "Two shoes times seven heavens equals one sun four doors. Two times seven equals fourteen.**

**Student writes: 2 x 7 = 14**

**2 x 8 = 16**

**Student draws the following pictures:**

**Student states: "Two shoes times eight gates equals one sun six sticks. Two times eight equals sixteen."**

**Student writes: 2 x 8 = 16**

**2 x 9 = 18**

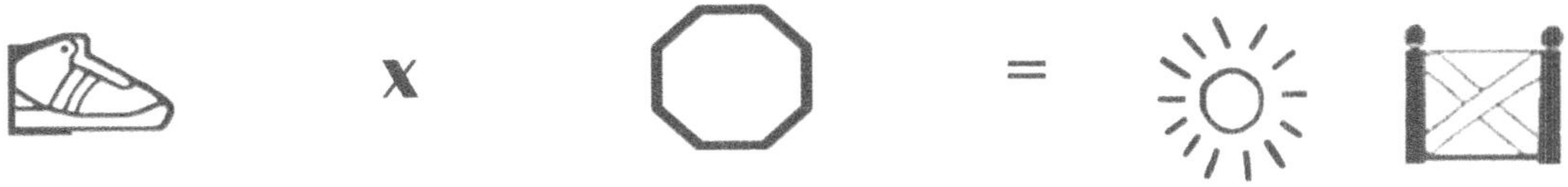

**Student draws the following pictures:**

**Student states: "Two shoes times nine signs equals one sun eight gates. Two times nine equals eighteen."**

**Student writes: 2 x 9 = 18**

**2 x 10 = 20**

**Student draws the following pictures:**

**Student states: "Two shoes times ten pins equals two shoes zero. Two times ten equals twenty."**

**Student writes: 2 x 10 = 20**

# 3 Three

3 x 0 = 0

 x 0 = 0

**Student draws the following pictures:**

 x 0 = 0

**Student states: "Three trees times zero equals zero. Three times zero equals zero."**

**Student writes: 3 x 0 = 0**

| 3 | x | 1 | = | 3 |
|---|---|---|---|---|

**Student draws the following pictures:**

**Student states: "Three trees times one sun equals three trees. Three times one equals three."**

**Student writes: 3 x 1 = 3**

**3 x 2 = 6**

**Student draws the following pictures:**

**Student states: "Three trees times two shoes equals six sticks. Three times two equals six."**

**Student writes: 3 x 2 = 6**

3 x 3 = 9

**Student draws the following pictures:**

**Student states: "Three trees times three trees equals nine signs. Three times three equals nine."**

**Student writes: 3 x 3 = 9**

3 x 4 = 12

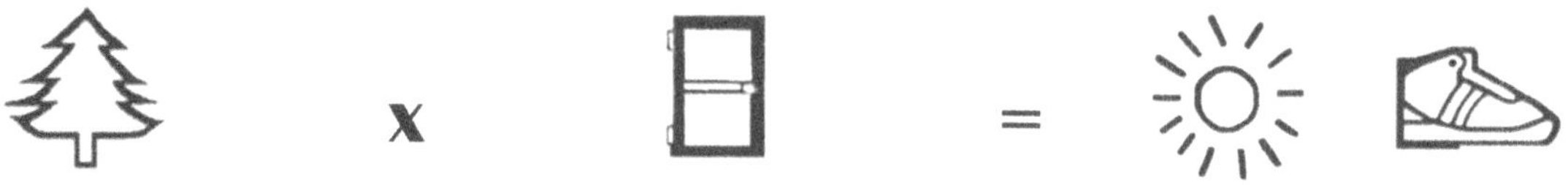

**Student draws the following pictures:**

**Student states: "Three trees times four doors equals one sun two shoes. Three times four equals twelve."**

**Student writes: 3 x 4 = 12**

3 x 5 = 15

**Student draws the following pictures:**

**Student states: "Three trees times five hives equals one sun five hives. Three times five equals fifteen."**

**Student writes: 3 x 5 = 15**

3 x 6 = 18

**Student draws the following pictures:**

**Student states: "Three trees times six sticks equals one sun eight gates. Three times six equals eighteen."**

**Student writes: 3 x 6 = 18**

**3 x 7 = 21**

**Student draws the following pictures:**

**Student states: "Three trees times seven heavens equals two shoes one sun. Three times seven equals twenty one."**

**Student writes: 3 x 7 = 21**

3 x 8 = 24

**Student draws the following pictures:**

**Student states: "Three trees times eight gates equals two shoes four doors. Three times eight equals twenty four."**

**Student writes: 3 x 8 = 24**

**3 x 9 = 27**

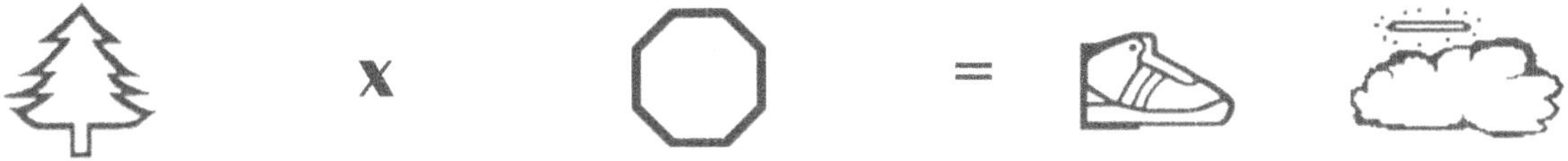

**Student draws the following pictures:**

**Student states: "Three trees times nine signs equals two shoes seven heavens. Three times nine equals twenty seven."**

**Student writes: 3 x 9 = 27**

3 x 10 = 30

 x  =  0

**Student draws the following pictures:**

 x  =  0

**Student states: "Three trees times ten pins equals three trees zero. Three times ten equals thirty."**

**Student writes: 3 x 10 = 30**

# 4 Four

**4 x 0 = 0**

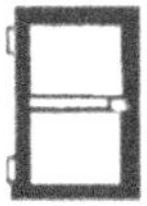 **x 0 = 0**

**Student draws the following pictures:**

 **x 0 = 0**

**Student states: "Four doors times zero equals zero. Four times zero equals zero."**

**Student writes: 4 x 0 = 0**

4 x 1 = 4

**Student draws the following pictures:**

**Student states: "Four doors times one sun equals four doors. Four times one equals four."**

**Student writes: 4 x 1 = 4**

4 x 2 = 8

**Student draws the following pictures:**

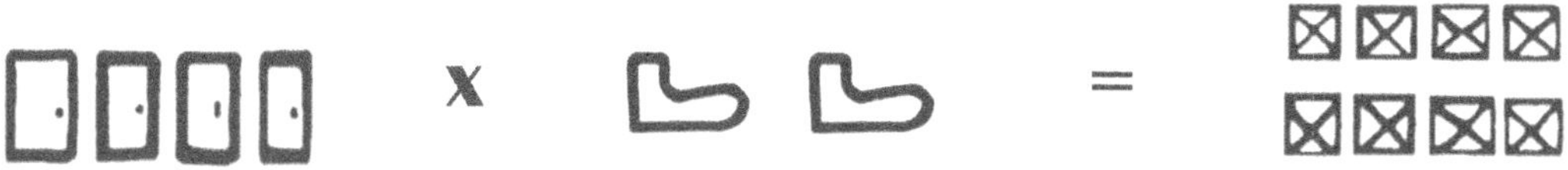

**Student states: "Four doors times two shoes equals eight gates. Four times two equals eight."**

**Student writes: 4 x 2 = 8**

4 x 3 = 12

**Student draws the following pictures:**

**Student states: "Four doors times three trees equals one sun two shoes. Four times three equals twelve."**

**Student writes: 4 x 4 = 12**

4 x 4 = 16

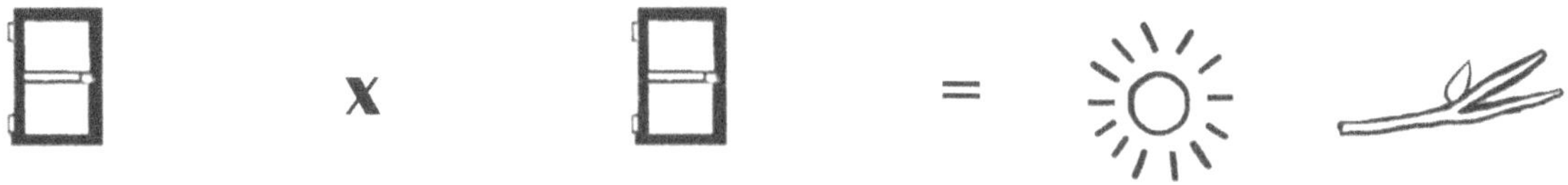

Student draws the following pictures:

Student states: "Four doors times four doors equals one sun six sticks. Four times four equals sixteen."

Student writes: 4 x 4 = 16

4 x 5 = 20

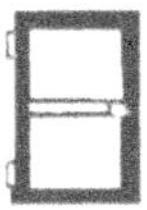 x  =  0

**Student draws the following pictures:**

 x  =  0

**Student states: "Four doors times five hives equals two shoes zero. Four times five equals twenty."**

**Student writes: 4 x 5 = 20**

4 x 6 = 24

Student draws the following pictures:

Student states: "Four doors times six sticks equals two shoes four doors. Four times six equals twenty four."

Student writes: 4 x 6 = 24

4 x 7 = 28

Student draws the following pictures:

Student states: “Four doors times seven heaven equals two shoes eight gates. Four times seven equals twenty eight.”

Student writes: 4 x 7 = 28

**4 x 8 = 32**

**Student draws the following pictures:**

**Student states: "Four doors times eight gates equals three trees two shoes. Four times eight equals thirty two."**

**Student writes: 4 x 8 = 32**

4 x 9 = 36

**Student draws the following pictures:**

**Student states: "Four doors times nine signs equals three trees six sticks. Four times nine equals thirty six."**

**Student writes: 4 x 9 = 36**

4 x 10 = 40

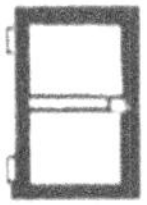 x  = 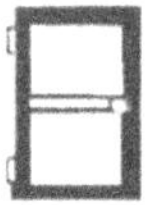 0

**Student draws the following pictures:**

 x  =  0

**Student states: "Four doors times ten pins equals four doors zero. Four times ten equals forty."**

**Student writes: 4 x 10 = 40**

5 Five

5 x 0 = 0

 x 0 = 0

**Student draws the following pictures:**

 x 0 = 0

**Student states: "Five hives times zero equals zero. Five hives times zero equals zero."**

**Student writes: 5 x 0 = 0**

5 x 1 = 5

 x  = 

**Student draws the following pictures:**

 x  = 

**Student states: " Five hives times one sun equals five hives. Five times one equals five."**

**Student writes: 5 x 1 = 5**

5 x 2 = 10

 x  = 

**Student draws the following pictures:**

 x  = 

**Student states: " Five hives times two shoes equals ten pins. Five times two equals ten."**

**Student writes: 5 x 2 = 10**

**5 x 3 = 15**

**Student draws the following pictures:**

**Student states: " Five hives times three trees equals one sun five hives. Five times three equals fifteen."**

**Student writes: 5 x 3 = 15**

5 x 4 = 20

 x 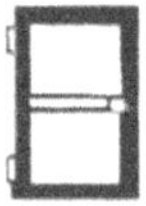 =  0

**Student draws the following pictures:**

 x  =  0

**Student states: " Five hives times four doors equals two shoes zero. Five times four equals twenty."**

**Student writes: 5 x 4 = 20**

**5 x 5 = 25**

**Student draws the following pictures:**

**Student states: " Five hives times five hives equals two shoes five hives. Five time five equals twenty five."**

**Student writes: 5 x 5 = 25**

5 x 6 = 30

 x  =  0

**Student draws the following pictures:**

 x  =  0

**Student states: " Five hives times six sticks equals three trees zero. Five times six equals thirty."**

**Student writes: 5 x 6 = 30**

5 x 7 = 35

 x 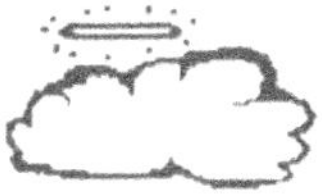 = 

**Student draws the following pictures:**

 x  = 

**Student states: " Five hives times seven heavens equals three trees five hives. Five times seven equals thirty five."**

**Student writes: 5 x 7 = 35**

**5 x 8 = 40**

 **x**  **=** 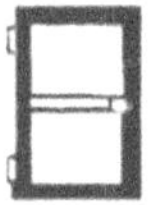 **0**

**Student draws the following pictures:**

 **x** 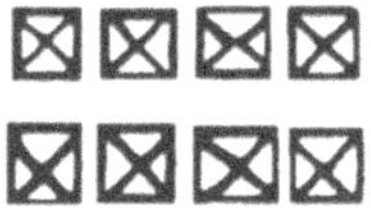 **=**  **0**

**Student states: " Five hives times eight gates equals four doors. Five times eight equals forty."**

**Student writes: 5 x 8 = 40**

5 x 9 = 45

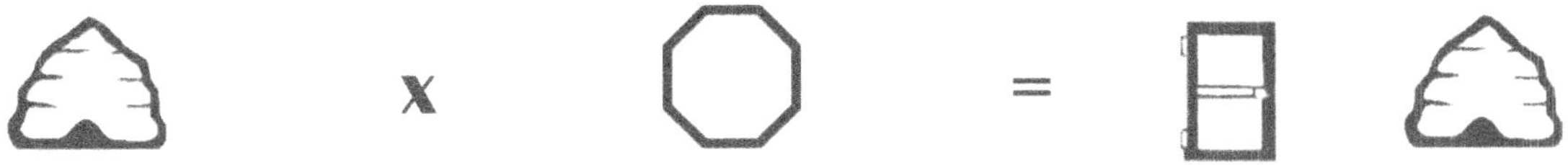

Student draws the following pictures:

Student states: " Five hives times nine signs equals four doors five hives. Five times nine equals forty five."

Student writes: 5 x 9 = 45

5 x 10 = 50

 x  =  0

**Student draws the following pictures:**

 x  =  0

**Student states: " Five hives times ten pins equals five hives zero. Five times ten equals fifty."**

**Student writes: 5 x 10 = 50**

# 6 Six

6 x 0 = 0

 x 0 = 0

**Student draws the following pictures:**

 x 0 = 0

**Student states: "Six sticks times zero equals zero. Six times zero equals zero."**

**Student writes: 6 x 0 = 0**

6 x 1 = 6

 x  = 

**Student draws the following pictures:**

 x  = 

**Student states: "Six sticks times one sun equals six sticks. Six times one equals six."**

**Student writes: 6 x 1 = 6**

6 x 2 = 12

Student draws the following pictures:

Student states: "Six sticks times two shoes equals one sun two shoes. Six times two equals twelve."

Student writes: 6 x 2 = 12

**6 x 3 = 18**

**Student draws the following pictures:**

**Student states: "Six sticks times three trees equals one sun eight gates. Six times three equals eighteen."**

**Student writes: 6 x 3 = 18**

6 x 4 = 24

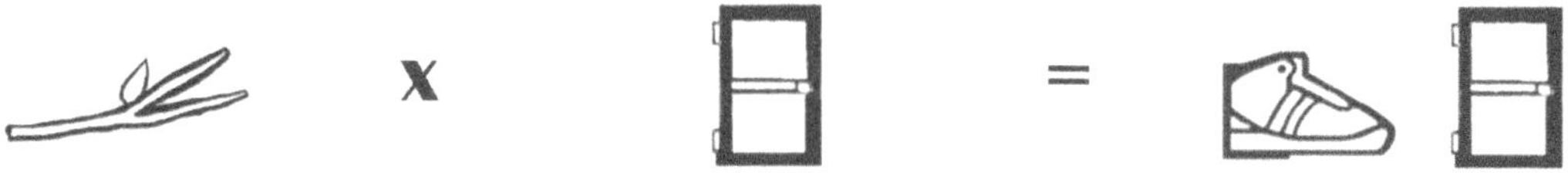

Student draws the following pictures:

Student states: "Six sticks times four doors equals two shoes four doors. Six times four equals twenty four."

Student writes: 6 x 4 = 24

**6 x 5 = 30**

 **x**  **=**  **0**

**Student draws the following pictures:**

 **x**  **=**  **0**

**Student states: "Six sticks times five hives equals three trees zero. Six times five equals thirty."**

**Student writes: 6 x 5 = 30**

**6 x 6 = 36**

**Student draws the following pictures:**

**Student states: "Six sticks times six sticks equals three trees six sticks. Six times six equals thirty six."**

**Student writes: 6 x 6 = 36**

6 x 7 = 42

**Student draws the following pictures:**

**Student states: "Six sticks times seven heavens equals four doors two shoes. Six times seven equals forty two."**

**Student writes: 6 x 7 = 42**

**6 x 8 = 48**

**Student draws the following pictures:**

**Student states: "Six sticks times eight gates equals four doors eight gates. Six times eight equals forty eight."**

**Student writes: 6 x 8 = 48**

6 x 9 = 54

 x 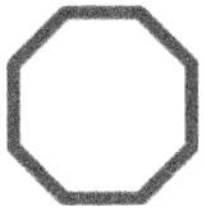 =  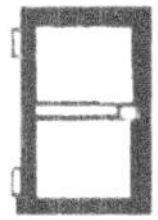

**Student draws the following pictures:**

 x  = 

**Student states: "Six sticks times nine signs equals five hives four doors. Six times nine equals fifty four."**

**Student writes: 6 x 9 = 54**

6 x 10 = 60

 x  =  0

**Student draws the following pictures:**

 x  =  0

**Student states: "Six sticks times ten pins equals six sticks zero. Six times ten equals sixty."**

**Student writes: 6 x 10 = 60**

# 7 Seven

7 x 0 = 0

 x 0 = 0

**Student draws the following pictures:**

 x 0 = 0

**Student states: "Seven heavens times zero equals zero. Seven times zero equals zero."**

**Student writes: 7 x 0 = 0**

7 x 1 = 7

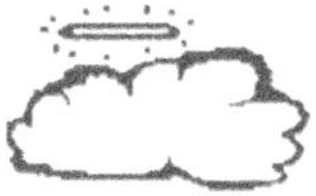 x  = 

**Student draws the following pictures:**

 x  = 

**Student states: "Seven heavens times one sun equals seven heavens. Seven times one equals seven."**

**Student writes: 7 x 1 = 7**

**7 x 2 = 14**

**Student draws the following pictures:**

**Student states: "Seven heavens times two shoes equals one sun four doors. Seven times two equals fourteen."**

**Student writes: 7 x 2 = 14**

7 x 3 = 21

**Student draws the following pictures:**

**Student states: "Seven heavens times three trees equals two shoes one sun. Seven times three equals twenty one."**

**Student writes: 7 x 3 = 21**

7 x 4 = 28

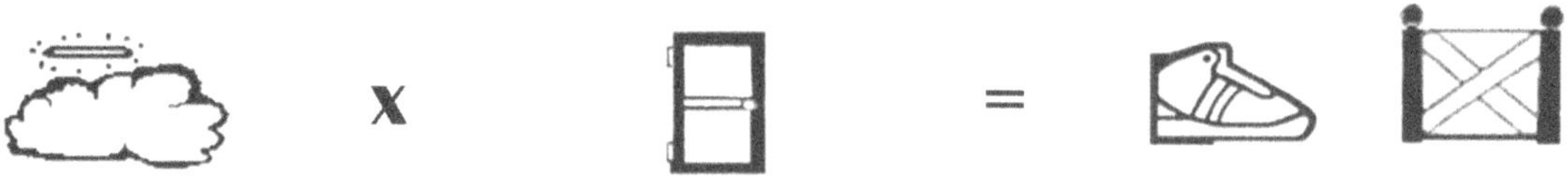

**Student draws the following pictures:**

**Student states: "Seven heavens times four doors equals two shoes eight gates. Seven times four equals twenty eight."**

**Student writes: 7 x 4 = 28**

**7 x 5 = 35**

**Student draws the following pictures:**

**Student states: "Seven heavens times five hives equals three trees five hives. Seven times five equals thirty five."**

**Student writes: 7 x 5 = 35**

7 x 6 = 42

**Student draws the following pictures:**

**Student states: "Seven heavens times six sticks equals four doors two shoes. Seven times six equals forty two."**

**Student writes: 7 x 6 = 42**

**7 x 7 = 49**

**Student draws the following pictures:**

**Student states: "Seven heavens times seven heavens equals four doors nine signs. Seven times seven equals forty nine"**

**Student writes: 7 x 7 = 49**

**7 x 8 = 56**

 x  =  

**Student draws the following pictures:**

 x 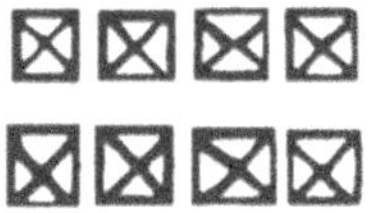 =  

**Student states: "Seven heavens times eight gates equals five hives six sticks. Seven times eight equals fifty six."**

**Student writes: 7 x 8 = 56**

**7 x 9 = 63**

 **x** 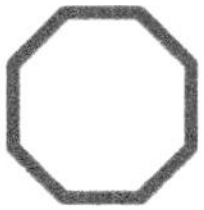 **=** 

**Student draws the following pictures:**

 **x**  **=** 

**Student states: "Seven heavens times nine signs equals six sticks three trees. Seven times nine equals sixty three."**

**Student writes: 7 x 9 = 63**

**7 x 10 = 70**

 **x**  **=**  **0**

**Student draws the following pictures:**

 **x**  **=**  **0**

**Student states: "Seven heavens times ten pins equals seven heavens zero. Seven times ten equals seventy."**

**Student writes: 7 x 10 = 70**

# 8 Eight

8 x 0 = 0

 x 0 = 0

Student draws the following pictures:

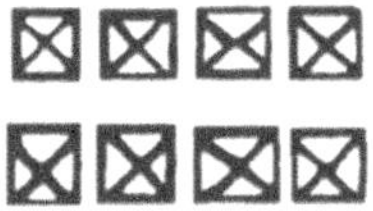 x 0 = 0

Student states: "Eight gates times zero equals zero. Eight times zero equals zero."

Student writes: 8 x 0 = 0

8 x 1 = 8

 x  = 

**Student draws the following pictures:**

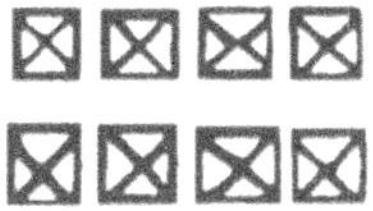 x  = 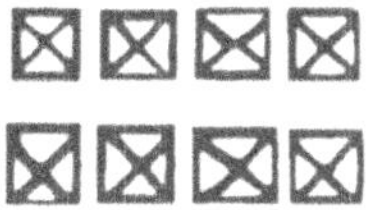

**Student states: "Eight gates times one sun equals eight gates. Eight times one equals eight."**

**Student writes:** 8 x 1 = 8

8 x 2 = 16

 x  = 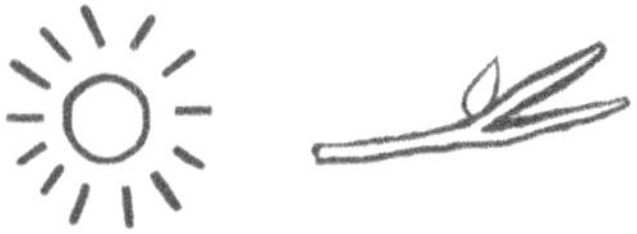

**Student draws the following pictures:**

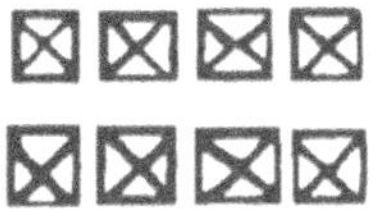 x  = 

**Student states: "Eight gates times two shoes equals one sun six sticks. Eight times two equals sixteen."**

**Student writes: 8 x 2 = 16**

**8 x 3 = 24**

**Student draws the following pictures:**

**Student states: "Eight gates times three trees equals two shoes four doors. Eight times three equals twenty four."**

**Student writes: 8 x 3 = 24**

8 x 4 = 32

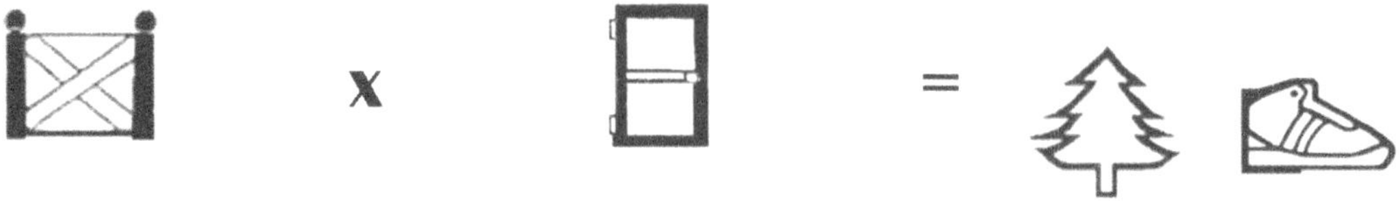

**Student draws the following pictures:**

**Student states: "Eight gates times four doors equals three trees two shoes. Eight times four equals thirty two."**

**Student writes: 8 x 4 = 32**

**8 x 5 = 40**

 **x**  **=** 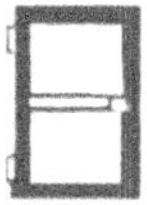 **0**

**Student draws the following pictures:**

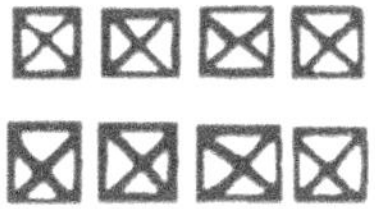 **x**  **=**  **0**

**Student states: "Eight gates times five hives equals four doors zero. Eight times five equals forty."**

**Student writes: 8 x 5 = 40**

8 x 6 = 48

 x  = 

**Student draws the following pictures:**

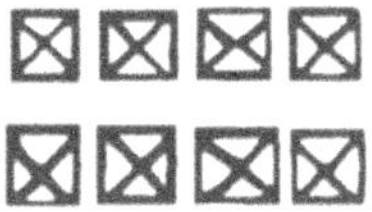 x  =  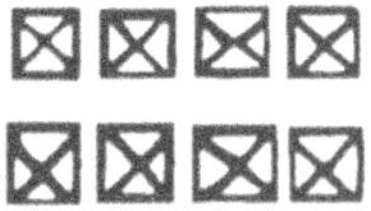

**Student states: "Eight gates times six sticks equals four doors eight gates. Eight times six equals forty eight."**

**Student writes:** 8 x 6 = 48

8 x 7 = 56

Student draws the following pictures:

Student states: "Eight gates times seven heavens equals five hives six sticks. Eight times seven equals fifty six"

Student writes: 8 x 7 = 56

8 x 8 = 64

 x  = 

**Student draws the following pictures:**

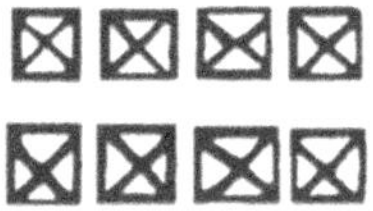 x 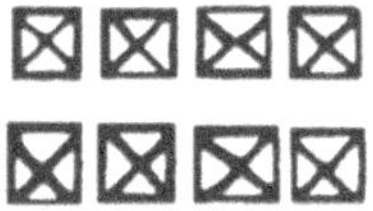 = 

**Student states: "Eight gates times eight gates equals six sticks four doors. Eight times eight equals sixty four."**

**Student writes: 8 x 8 = 64**

8 x 9 = 72

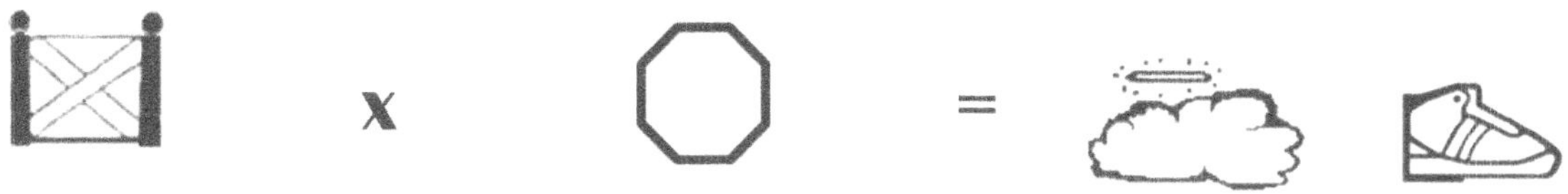

**Student draws the following pictures:**

**Student states: "Eight gates times nine signs equals seven heavens two shoes. Eight times nine equals seventy two."**

**Student writes: 8 x 9 = 72**

8 x 10 = 80

 x  =  0

Student draws the following pictures:

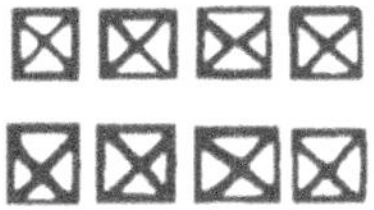 x  = 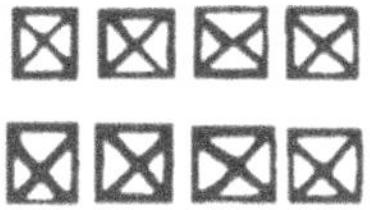 0

Student states: "Eight gates times ten pins equals eight gates zero. Eight times ten equals eighty."

Student writes: 8 x 10 = 80

# 9 Nine

9 x 0 = 0

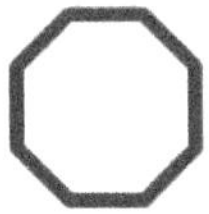 x 0 = 0

**Student draws the following pictures:**

 x 0 = 0

**Student states: "Nine signs times zero equals zero. Nine times zero equals zero."**

**Student writes: 9 x 0 = 0**

9 x 1 = 9

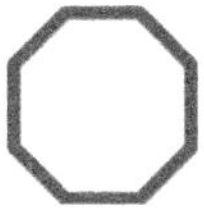 x  = 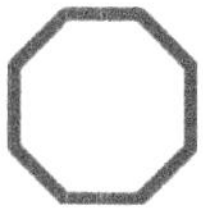

**Student draws the following pictures:**

 x  = 

**Student states: "Nine signs times one sun equals nine signs. Nine times one equals nine."**

**Student writes: 9 x 1 = 9**

9 x 2 = 18

**Student draws the following pictures:**

**Student states: "Nine signs times two shoes equals one sun eight gates. Nine times two equals eighteen."**

**Student writes: 9 x 2 = 18**

# 9 x 3 = 27

**Student draws the following pictures:**

**Student states: "Nine signs times three trees equals shoes seven heavens. Nine times three equals twenty seven."**

**Student writes: 9 x 3 = 27**

9 x 4 = 36

**Student draws the following pictures:**

**Student states: "Nine signs times four doors equals three trees six sticks. Nine times four equals thirty six."**

**Student writes: 9 x 4 = 36**

**9 x 5 = 45**

**Student draws the following pictures:**

**Student states: "Nine signs times five hives equals four doors five hives. Nine times five equals forty five."**

**Student writes: 9 x 5 = 45**

9 x 6 = 54

**Student draws the following pictures:**

**Student states: "Nine signs times six sticks equals five hives four doors. Nine times six equals fifty four."**

**Student writes: 9 x 6 = 54**

9 x 7 = 63

**Student draws the following pictures:**

**Student states: "Nine signs times seven heavens equals six sticks three trees. Nine times seven equals sixty three."**

**Student writes: 9 x 7 = 63**

**9 x 8 = 72**

**Student draws the following pictures:**

**Student states: "Nine signs times eight gates equals seven heavens two shoes. Nine times eight equals seventy two."**

**Student writes: 9 x 8 = 72**

**9 x 9 = 81**

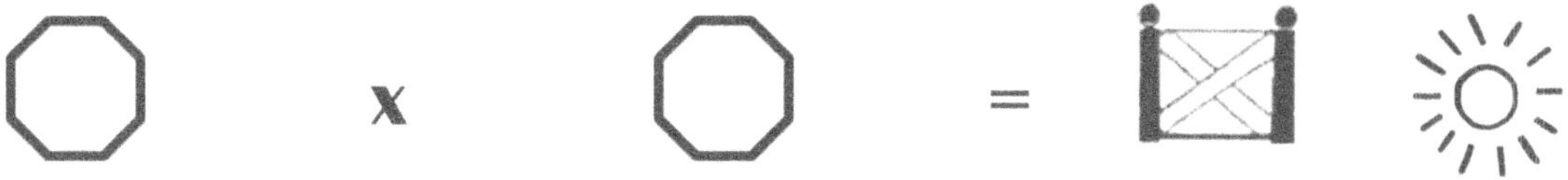

**Student draws the following pictures:**

**Student states: "Nine signs times nine signs equals eight gates one sun. Nine times nine equals eighty one."**

**Student writes: 9 x 9 = 81**

9 x 10 = 90

 x  = 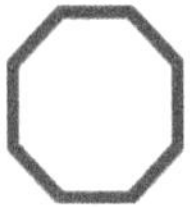 0

**Student draws the following pictures:**

 x  =  0

**Student states: "Nine signs times ten pins equals nine signs zero. Nine times ten equals ninety."**

**Student writes: 9 x 10 = 90**

Made in United States
Cleveland, OH
04 November 2026

47647259R00065